# How To Grow Healthy & Tasty Cucumbers

I0482565

## *Quick Start Guide*

HTeBooks

**Disclaimer**

This book is designed to provide condensed information. It is not intended to reprint all the information that is otherwise available, but instead to complement, amplify and supplement other texts. You are urged to read all the available material, learn as much as possible and tailor the information to your individual needs.

Every effort has been made to make this book as complete and as accurate as possible. However, there may be mistakes, both typographical and in content. Therefore, this text should be used only as a general guide and not as the ultimate source of information. The purpose of this book is to educate.

The author or the publisher shall have neither liability nor responsibility to any person or entity regarding any loss or damage caused, or alleged to have been caused, directly or indirectly, by the information contained in this book.

# Table of Contents

# How Will This Book Help You?

Growing a plant, any kind of plant for that matter, is an art in itself. Growing cucumbers is especially so. It not only is tasking to attempt to grow healthy cucumbers that are tasty and fast selling; it is a lot of fun as well. This book aims to help you combine both functionality and the element of fun as you go about growing your cucumbers. This book aims to provide you with the right information to grow cucumbers that are healthy and tasty, given whatever tools you have at hand.

This is a direct book by all means, structured in a way that helps you grasp tips and useful bits while taking as little time as possible. Since it is written with the intention to aid all sorts of persons, it will also simplify the processes and elements to input in a basic, pared down way.

The cucumber, while not recognized as such, could well be the nation's plant of pride, just as the clover is to the Celts. Read on to pick up the finest tips available to help you own the most attractive cucumber garden around.

# Growing Cucumbers: Basics To Know When Starting Out & Soil, Planting And Care

*"The beginnings are everything; take them seriously"*

*- Hugh Grant*

## Part I: Growing Cucumbers: Basics to Know when starting out

The first thing you need to know about cucumbers is that they are tropical plants. They tend to do best when temperatures are high and water is in plenty. This being said, growing cucumbers is meant for the warmer weather. Cucumber plants are so tender, with regard to frost, that you must not set them in the garden until the soil temperatures have reached the reliable range (that is the 70 degrees range). Even when you have ascertained the soil temperatures to be safely in the 70s, absolutely desist from setting your plants in the soil until two weeks after the last recorded frost date.

Your cucumber plants will take on two growth forms: bush and vining. The vines will grow along the ground or climb up trellises if they are available. The bush kind (such as the bush hybrids) will generally take on a more compact, space-friendly form. You must understand that in general, the vining cucumbers tend to have a higher degree of produce, compared to the bush kind. However if you are strapped for space, or are into indoor gardening, the bush kind will suit your space needs best as they are better suited. You may boost the overall yield of the season (strictly with regard to the bush cucumber kind) by planting a number of cucumber crops in close succession (about two weeks apart).

Whether your cucumber planting is geared towards slicing or simply picking, you are assured of getting enough variety to make your pick from. The lemon cucumber will give you smaller fruits that are just perfect for a singular serving while the Boston picking kind has a classic heirloom taste to it that is just phenomenal. The long Armenian cucumber is an ethnic cucumber that has been pried for years for its great taste as well as the fact that one cucumber provides a numerous amount of slices. With cucumber choices, it is impossible not to find the right kind for you.

## Part II: Soil, cucumber planting and overall care

Of course, cucumbers will need fertile soils if they are to grow healthy as well as tasty. But the soils need to be warm too, as the subchapter above stresses. The pH needs to be slightly acidic (in the region of 6.5). However, there are allowances for acidity as high as 6.0 or as low as 6.8. Going past these however is taking unnecessary risks.

## Tips on what to do when planting your cucumber plants

### *Planting basics*

The first step is composting. You need to work some composted manure into your soil (for fertility purposes).

Once this is done, get good quality seedlings. Take care to plant your seedlings 36-60 inches apart. This will be fully reliant on the variety you have. It is equally helpful to refer to the tag. There will always be one.

For the vine kind that is trained on trellises, space your plants about 1 foot apart.

If you are in an area where spring season is both cool and long, you should warm your soil to about 4 degrees by covering up your hill or row with some black plastic. If you prefer to keep off black plastic in your planting endeavors, you should mulch with any pine straw, chopped leaves or wheat straw, just to name a few.

### Basics on taking care of your young cucumbers

If the weather is cool to the point where you consider it to be unseasonably cool, you really ought to wait until you are sure the soil has been sufficiently warmed by the sun.

Mulch will be very important if you are to keep your cucumbers clean (usually in the case of bush type cucumbers or those vining kinds that you are not training on a trellis).

Straw mulch is immensely uncomfortable for slugs as well as providing some very unsure footing for cucumber beetles. If you can access straw mulch, it should be your first priority.

If it is possible to accomplish, take time to trellis your vines. This will not only save you a lot of space, it will also aid you in your endeavors to keep your cucumbers clean. Here are some specific measurements you may keep as standards: a 15 inch diameter cage that has been fashioned from 4.5 foot (or so) welded fencing wire or even hog wire can support up to 3 vines. Wire makes the climbing efforts of the cucumber tendrils as the plant grows easier.

Cucumbers grow fast and at the same time, do not demand a lot of care. All you need to do is keep your soil consistently moist with about an inch of water weekly. If the temperatures are really high, increase that amount of water. Do the same if rains are especially scarce.

If it is possible, you should opt to use a soaker hose for your watering efforts. Drip irrigation also produces desirable results. This kind of watering also helps keep leaf diseases at bay.

You may fertilize your cucumbers with liquid food. A good example is the Bonnie range of liquid plant food. Use the liquid food about twice every month (you may do your applications after every two weeks for consistency). Apply the liquid food directly in the soil that is around the stems of the plant.

Granular, slow-release fertilizer is another option. Work it into the soil after you do your planting or simply sprinkle it around the plants later.

**\*Key point/action step**
If you want the maximum yield for your cucumber garden, you must set up a strong foundation. The foundation here means ensuring

that you provide the right environmental conditions and spacing the cucumbers properly.

# Step By Step Actions To Take For Healthy & Tasty Cucumbers: Comprehensive Steps For Planting, Taking Care & Dealing With Pests

*"Without the trunk, how can you have anything? Only a fool ignores the middle parts, regardless of how unspectacular they seem."*

*-Anon*

Cucumbers are veggies that thrive best in warm weather. Before you delve into the finer bits of planting and taking full care of your cucumber patch, you must keep in mind to plant them at least 2 weeks after the last frost date, lest they fall victim to the harmful effects of frost. Especially being that they are immensely frost sensitive, it is particularly important to keep this in mind.

Most cucumber varieties are not choosy- they will happily grow in any space, especially the vining kind (attributable to its ability to climb). The most common slicing cucumber varieties will have sprawling vines that have large-sized, green leaves and tendrils that curl. Cucumbers grow at quite some speed and if you follow the directives that this chapter provides, your returns will be consistently abundant.

## Planting your cucumbers

Cucumbers must be introduced into the soil no earlier than two weeks after the last recorded frost date. This is owing to the fact that cucumbers are very sensitive to frost and as such, are very susceptible to the damage that comes along with frost. Your soils must be at least 65 degrees Fahrenheit before you do any planting, though the recommendation is to have soil temperatures in the 70s.

In order to have an early crop, start your seeds indoors some three weeks before you go about transplanting them in the soil. Cucumbers prefer a bottom heat of around 70 degrees Fahrenheit (around 21 degrees Celsius). If at all, a heat mat is not available, place your seeds flat on top of your refrigerator, or place a few of them on top of your water heater. It will go a long way in helping out.

Before you go about planting, select a spot for planting. Basically, an ideal spot is the site that gets a generous amount of sun and preferably, is shade-free.

Ideally, the soils should be neutral, slightly acidic or slightly alkaline. You can improve your clay soil by adding in composted manure (to boost the fertility). If you only have dense soils that are heavy, you can improve them by adding peat to them. Rotted manure or compost will also do an equally good job. If you are unsure about the kind of soil that you have to deal with, it will do you good to have a soil test done. In the case of Northern gardens, light, sandy soils are preferred. This is because they tend to warm more quickly, compared to the other soils.

Mix in some compost or aged manure in your soil then plant about 2 inches deep. Proceed to work into the soil about 6 to eight inches in depth. Be sure to make sure that your soil is well drained and moist. Soggy soils will only reduce your chances of growing some fine, healthy and tasty cucumbers.

Sow your seeds in neat rows. Let the depth be roughly 1 inch deep and the spacing be 10 inches apart.

If at all you are transplanting your seedlings, you will need to be a bit more liberal with your spacing. Allow a 12-inch space between individual seedlings.

If you want your vine to climb (which is what you should want), a trellis is a phenomenal idea. Have one around for this purpose. The trellis will save you a lot of growth space. In addition, a trellis will protect your cucumbers from damage caused by the cucumber fruit lying on moist ground.

**Caring for your cucumbers**

When you are planting your seeds into the ground, take care to cover with some netting material. As a substitute, you may use a berry basket. The purpose of this is to keep off pests from digging out the seeds to feed on.

Immediately your cucumber seedlings emerge, begin to water them. Keep your watering sessions frequent. After you observe the formation of fruit, increase the amount of water you use for water to about a gallon a week.

Observe the growth of your seedlings keenly. Once they have reached 4 inches in height, thin them. You will now observe a one and a half feet space between individual plants. This is what you should aim for.

If you took the pains to add in organic matter to your soil before you set about planting your cucumbers, you may only require to side dress your plants with some well-rotted manure or some compost. If you so wish, you can opt for a fertilizer from your local store. Select one that is low in Nitrogen levels but high in potassium and phosphorous. Apply this at planting, exactly one week after blooming and every three weeks with some quality liquid food. Target the soil that is around the plant and apply directly to it. Over-fertilizing will only get your plants stunted, so avoid it.

Water your plants consistently. You are not going to raise healthy cucumbers that are also tasty if you do not water your plants religiously. If anything, chances are you will not get to enjoy the fruits of your efforts if you do not take watering seriously. Stick your finger into the soil. If you notice dryness past the first joint of your finger, then by all means understand that your plants need to be watered. You should also know that inconsistent watering will only lead to fruit that has a bitter taste to it. Let your watering sessions be especially slow: water in the morning and afternoon. Always take care to keep the watering hose away from the plant leaves.

You should mulch. Mulching is beneficial in that it helps keep in the soil moisture.

In the case of limited space or you simply want vines that are vertical, set up trellises early. This is to avoid any damage to the seedlings as well as the vines.

Take time to spray your vines with sugar water. This will attract bees and allow the setting of a higher number of fruit.

## Fruiting problems & dealing with pests

To begin with, understand that your cucumbers may not set fruit if at all the first flowers were all male flowers. You should know that both male and female flowers must be growing at the same period. This may well not happen early in your plants' life. So what is the solution? Simply be patient and sit tight. Eventually, the gender balance will naturally establish itself and you will be able to enjoy observing your first fruit emerge.

The lack of fruit may also be attributed to poor pollination by bees. This is especially so if the bees are prevented from effective pollination by factors such as insecticides, cold temperatures and rain. You must keep in mind that gynoecious hybrids must have pollinator plants.

### *Dealing with cucumber beetles:*

What are cucumber beetles? If you discover that the stems of your seedlings are being bitten off, your cucumber leaves are yellowing and wilting and there is the presence of holes, then you may have stripped cucumber beetles to deal with.

Often, these beetles leave their sites of hibernation early on in the season to feed on your seedlings as they emerge (they are strongly attracted by cucurbit veggies. Cucumbers are, of course, a member of these. Others include pumpkins, squash and beans). Their attacks on your seedlings will often end up in your plants dying early on. This is especially so when the ravages of the cucumber beetles combine with the destructive action of the beetle larvae on the roots of your seedlings.

## Controlling cucumber beetles:

Inspect your newly planted cucumbers for the presence of this beetle. Especially when your plants are still seedlings, be very watchful.

Cover your seedlings with row covers. However, you must remember to remove these for a few hours every day during blossoming time to allow for pollination.

Tilling your garden in late fall will expose any cucumber beetles hiding there to the harsh conditions of winter and thus, cut down on their population the next year. Tilling will also make your soils easier to work on in spring.

Use natural predators for these beetles. A good example is soldier beetles. You could also opt for braconid wasps as well as some nematodes.

## Dealing with white flies:

Spray your cucumbers with some choice insecticidal soap. Follow up this process a couple of times, or even three.

Cucumbers are especially sensitive to insecticides. If you can introduce spiders and ladybugs among your plants, this will be especially good to combat white flies. They serve as a very potent control for these flies.

Try this mixture to put off whiteflies and control them: in a spray bottle, mix some rubbing alcohol (two parts of it) with 5 parts of water and a tablespoon of liquid soap. Spray this mixture on the cucumber foliage, targeting those that you suspect to be under the ravages of whiteflies.

**\*Key point/action step**

You need to put in a lot of effort to grow healthy and tasty cucumbers. You need to put your focus on mulching, fighting white flies, and cucumber beetles if you really want to have the best yield. You should as well take care to maintain the right pH as well as provide all essential nutrients for maximum growth.

# Bush Cucumbers Versus Vining Cucumbers: What You Need To Know About Growing Both Kinds So As To Know Which To Choose For The Best Quality Fruits

*"There are no ghosts, save for them that come back to remind you of some poor choice. So choose wisely…every time"*

*- Soto*

Gardeners often have many choices, especially when it comes to veggies. With cucumbers, this is no different. Actually, even that cucumber that an adept farmer or consumer may consider lowly comes in hundreds upon hundreds of unique varieties that have been bred for use either as slicers or as picklers. With the modern cucumber plants, growth is often conducted to include as few male flowers as possible. This has the effect of increasing the produce. Little wonder then, that some individual would have hooked his attention onto the bush kind cucumber and endorsed it as a great garden plant even when the vining cucumber was still in widespread use.

Here is what you need to know about both kinds before you make up your choice on; your cucumber of choice.

## The bush cucumber

Primarily, these are bred for their phenomenal space saving qualities, seeing as they use up very little space with their very short vines. Most varieties will only call for a maximum of 3 square feet per plant.

The cultural requirements are similar to those of the vining cucumber and they mature and ripen at just about the same time and rate. Examples of popular bush cucumber varieties include the bush champion, pickle bush, parks bush whopper, salad bush, potluck and space-master.

## The advantages of bush cucumbers

If you are a fan of indoor gardening, then the bush cucumber is what you should opt for. It is ideal for indoor container gardening as well as small gardens.

Bush cucumbers are not very fussy with where you plant them, though you must ensure that you provide optimum growth conditions for your plant. However, they will grow phenomenally in just about any well-drained space you plant them in, in your garden. Especially if the air circulation is good, you can be assured of impressive returns.

For their size, the amounts of produce that they give you are just phenomenal. However, they will also not overwhelm you with produce that is far too abundant for your use and consumption.

Basically, if you have a small family or are simply not that keen with a bumper crop of cucumber fruits, choose the bush type. It is space friendly quality and this makes it all the more ideal.

## Vining cucumbers

Vining cucumbers, especially if given the license to roam, use up quite the amount of garden space. When they are trellised however, they not only make great use of available space, but they may also be used as landscaping screens. Such varieties like burpee hybrid, dasher 11, country fair 83, slice master, saladin, sweet success, sweet slice or slice nice are immensely popular choices with regard to vining cucumbers.

## The advantages of vining cucumbers

Although vining cucumbers will often demand for more planting space compared with the bush kind, they have been around for a longer time. As such, they have been bred into a much wider range of sizes and shapes. There exists no match, at least with regard to the bushing varieties, for such types as the lemon cucumber or the multiple white-skinned varieties of cucumber out there.

The vining kind often produces amazing amounts of fruit, especially when you compare them to the bush variety. This leaves you with ample extras for trading to friends, picking, and the like.

## *Key point/action step
Each type of cucumber requires a unique approach if you are to get the most output from your garden. You should therefore

understand the benefits that come with growing each type of cucumber before you start.

# Trouble Shooting, With Regard To Healthy Cucumbers, Harvesting & Storing Your Cucumbers For Best Preservation Of Quality

*"Just knowing what the problem is is solving it halfway."*

*- Allen Ginsberg*

## Troubleshooting, with regard to healthy & tasty cucumbers

If you discover that your vines are blooming but there is a marked absence of fruit, chances are that there is something getting in the way of effective pollination. First of all, make sure that you observe both the male and the female blooms. The male blooms will be the earliest to appear, before dropping off. If this happens, there is no cause for alarm. Within a couple of weeks at most, female flowers will appear, each one with a small swelling at the base, shaped like a cucumber. This is what later develops into a cucumber.

There are several pests that are a bother to cucumbers. Squash bugs have a very defined affinity for cucumber seedlings. The slug family on the other hand waits until the fruit has formed and is ripening before moving in. The aphids love to colonize not just the leaves, but the buds as well. Straw mulch gives slugs a very uncomfortable time, so you should opt for it, as well as setting up trellises which lift the fruit off the ground.

Cucumber beetles also bother vines to a large degree, chewing holes in the leaves as well as the flowers and leaving deep scars in the stems and fruits. Worse than this however, they spread a disease that causes the cucumber plant to wilt off and die. Powdery mildew is a cucumber disease that leaves white patches on the cucumber leaves. At the first sign of its presence, comence on the application of fungicides.

So as to cut down any spread of disease, do not harvest your crop when the leaves and vines are still wet.

### Harvesting and storage

Whenever your cucumbers are ripe and big enough for consumption, you may go about picking them.

Check the vines every day as the cucumber fruits start to appear because they tend to enlarge very quickly. The more you harvest, the more fruit the vines produce.

To remove the fruit cleanly, use some clippers or a knife, taking care to cut the stem just above the fruit. Tugging at the fruit will only leave you with a damaged vine.

Do not allow the cucumber fruits to get too big. The reason for this is that they become bitter tasting. They will also keep the vine from giving more produce.

Yellowing at the bottom of your cucumber is a signal for overipeness. You must remove the fruit immediately.

Harvest your lemon cucumbers just before they turn yellow. Although a major reason why they are called lemon cucumbers is because they turn yellow when ripening and end up looking a lot like a lemon (the shape contributes to this too), allowing the fruit to turn yellow may result in a taste that is a little too seedy for most people's tastes.

Keep those cucumbers that you harvest in the refrigerator for a period of 7 to 10 days. However, the recommendation is to use them as soon as you pick them. This is to get the flavor while it is at its best.

If you do not eat a slicing cucumber all in one sitting, wrap up what remains in plastic wrapping so as to prevent dehydration. Store in a refridgerator.

**\*Key point/action step**
You need to follow specific techniques to determine if your cucumbers have the best yield. You will also need to be aware of the best time to harvest if you really want to get the tastiest cucumbers from your garden.

# The Top 21 Must-Know Tips For Healthy & Tasty Cucumbers That Your Competition Is Probably Unfamiliar With

*"Any extra thing that keeps you above your competition, no matter how small it is, embrace and own."*

*- Napoleon Hill*

Nearly one half of the nation's veggie growers, roughly 47 percent for specifics, plant cucumbers according to one Susan Littlefield, the National Gardening Association Horticultural editor. This effectively makes cucumbers the number two most popular veggie that is homegrown. It should surprise no one that tomatoes are the top grown veggie, at a whopping 86 percent. Still, the cucumber percentage value is immensely impressive.

If you have a garden space that gets maximum sunshine, then growing your cucumbers will be an easy enough practice. Especially if you follow all of these unique directives in this chapter and are not a victim of late freezes in the spring, you should begin harvesting superbly sized and tasty cucumbers in no time.

## Planning and preparation tips

Always go for the disease resistant types. This is for obvious reasons- these will more effectively combat diseases thus boosting your chances of having a superb yield at the end of the day.

Always go for a fertile space that gets lots of sunshine every day. Cucumbers will do poorly in infertile soils. They may not make it at all if they are grown in places that are cold and receive no sunshine.

In order to have an earlier harvest as well as to greatly minimize the prospect of insect damage to your seedlings, start several plants indoors in individual pots. Trays with individual compartments are also a great substitute.

Set up several trellises or even a fence if you go for the vining kind. Trellises made of wire are best, as they make it easier for tendrils to wrap themselves around as the plant grows.

## Planting tips

Only sow your seeds in your outdoors garden after the danger of frost has well and truly passed away, and are sure that come what may, the soil will retain sufficient warmth for optimum growth. Cucumber plants are extremely susceptible to frost damage.

Make your second sowing some 5 weeks later to get a late summer or early fall harvest. This is a good way to manipulate harvest timelines.

When you are seeding in rows, keep them neat. Plant your seeds about 1 inch deep and some 6 inches apart.

When you are seeding in hills do this: plant four seeds in one-foot sized diameter circles. Set them some five or six feet apart.

## Caring for the cucumbers

When your plants are about 3 to 4 inches tall, start thinning them. Of course, this will depend on the type that you are working with (either pickling or slicing).

When growing your plants in hills, this is how to thin them: thin your plants to the healthiest two plants, in the phenomenon of plants having two or three leaves.

You must keep your soils evenly moist. Why is this? The reason is to keep your cucumbers from becoming bitter in taste.

About 4 weeks after you plant them in the soil, go about side dressing them. For very plant, apply two ample handfuls of compost. For each plant, keep the compost bands narrow.

After applying the fertilizer, apply a thick layer of mulch. Mulch will help keep the soil moisture locked into the soil, thus greatly benefitting the cucumber plants.

## Tips on controlling pests and diseases

Be very keen in monitoring your cucumbers as well as other veggies that are in close proximity to them for any buildup of insect pests.

It could be that the most effective way for the home gardener to control pests, with the example of the very destructive cucumber beetle high up the list, is to involve habits that shake up the life cycle of these insects as well as their habits. These will include you covering your young plants with some lightweight row covers up until flowering sets in. Also, exercise crop rotation- it is very effective in this.

If you do decide that pesticides and insecticides are the way to go, try to stick to the natural as much as you can. The less toxic the insecticide, the better it will be to use. Now, the only trouble you may perhaps face with this is that cucumber beetles are a hardy lot. There are not many effective natural, non-toxic insecticides available to deal with them.

You might have heard of Kaolin clay. If you have not, then this book will shed some light on it for you. Kaolin clay is perhaps the most effective natural insecticide choice to deal with cucumber beetles. It acts as a potent, long lasting repellent to them.

There is a big problem with the usage of broad-spectrum kind contact pesticides (these include malathion, cyhalothrin, permethrin, carbaryl and pyrethrin). The problem with these is that they not only kill the pests, they also kill off the beneficial predators as well as the parasites of the insect pests.

Insecticides are pretty much something you may have to resign yourself to using. So here is what to do with regard to them: read all the package labels keenly. For example, be familiar with what they advice about application and, say, harvesting. Is it a stipulation to wait for a few days after application before you harvest? These are the kinds of things you are looking to know.

You ought to consider capturing the pest that is ravaging your cucumbers. However, this is not to be used as a substitute for pesticides or natural control- it is far too impractical to work anyway. Rather, capture the insect or pest, lace it in a sealed bag, and then take it to the local garden center. Ask the staff to have a look at it and then give you any useful advice they have on the best control method in your particular area.

## Harvesting your cucumber plants

Harvest your cucumbers once they hit slicing or pickling size. Actually, do your harvesting every two days to prevent the probability of cucumber fruits from achieving an excessively large size as well as keep the plant consistently productive.

## *Key point/action step

Growing cucumbers entails different aspects that you need to master if you are to get the best output. Right from selecting the type of cucumber to grow through preparing the soil, taking care of the cucumbers as they grow, keeping off pests and harvesting, you must be careful if you really want to get the most output.

# How To Apply What You Have Learned?

This book has pretty much laid everything out as clearly and as straightforwardly as is possible. Most of the advice offered here is direct and comes with its own set of whys. So how do you go about using this book? This is easy enough. Simply exercise a much more practical approach to your cucumber tending compared to a more theoretical one. The book is very practical, often giving specific measurements, instructions, and directions. Thus, it will be of little help to you if you do not pick your patch out and set about planting your cucumbers.

This book directs you on growing the healthiest and tastiest cucumbers. Every directive is set to help you get the best. For the best results, try to keep within the confines given in this book to the highest degree that you can. This way, you can enjoy a superb produce that surprises even you.

www.ingramcontent.com/pod-product-compliance
Lightning Source LLC
Chambersburg PA
CBHW070303190526
45169CB00004B/1514